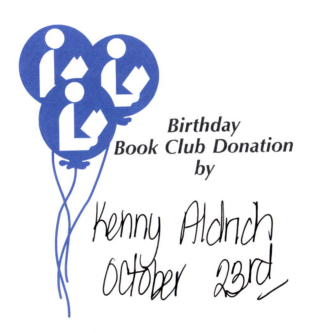

Birthday
Book Club Donation
by

Kenny Aldrich
October 23rd

© DEMCO, INC. 1990 PRINTED IN U.S.A.

Baby
Goats

Mary Elizabeth Salzmann

Consulting Editor, Diane Craig, M.A./Reading Specialist

Sandcastle

An Imprint of Abdo Publishing
www.abdopublishing.com

www.abdopublishing.com

Published by Abdo Publishing, a division of ABDO, PO Box 398166, Minneapolis, Minnesota 55439. Copyright © 2015 by Abdo Consulting Group, Inc. International copyrights reserved in all countries. No part of this book may be reproduced in any form without written permission from the publisher. SandCastle™ is a trademark and logo of Abdo Publishing.

Printed in the United States of America, North Mankato, Minnesota

102014
012015

Editor: Alex Kuskowski
Content Developer: Nancy Tuminelly
Cover and Interior Design and Production: Mighty Media, Inc.
Photo Credits: Shutterstock

Library of Congress Cataloging-in-Publication Data

Salzmann, Mary Elizabeth, 1968- author.
  Baby goats / Mary Elizabeth Salzmann.
     pages cm. -- (Baby animals)
  Audience: Ages 4-9.
  ISBN 978-1-62403-509-8
1. Kids (Goats)--Juvenile literature. 2. Goats--Juvenile literature. I. Title.
  SF383.35.S25 2015
  636.3'907--dc23
                        2014023425

SandCastle™ Level: Beginning

SandCastle™ books are created by a team of professional educators, reading specialists, and content developers around five essential components—phonemic awareness, phonics, vocabulary, text comprehension, and fluency—to assist young readers as they develop reading skills and strategies and increase their general knowledge. All books are written, reviewed, and leveled for guided reading, early reading intervention, and Accelerated Reader® programs for use in shared, guided, and independent reading and writing activities to support a balanced approach to literacy instruction. The SandCastle™ series has four levels that correspond to early literacy development. The levels are provided to help teachers and parents select appropriate books for young readers.

EMERGING • **BEGINNING** • TRANSITIONAL • FLUENT

Contents

Baby Goats 4

Did You Know? 22

Goat Quiz. 23

Glossary. 24

Baby Goats

Most baby goats are **twins**. These twin goats are white.

Baby goats can also be brown.

Some baby goats are several colors.

Baby goats get milk from their mothers.

Baby goats can be fed by people too.

Some baby goats live on farms.

Other baby goats live on mountains.

Baby goats like to climb on things.

Baby goats play together.

Did You Know?

- ▶ Baby goats are called kids.
- ▶ There are more than 300 goat **breeds**.
- ▶ Goats live for 15 to 18 years.
- ▶ Goats can be trained to pull carts.

Goat Quiz

Read each sentence below. Then decide whether it is true or false.

1. Baby goats are never **twins**.
2. Some baby goats are several colors.
3. Baby goats can be fed by people.
4. No baby goats live on farms.
5. Baby goats don't like to climb on things.

Answers: 1. False 2. True 3. True 4. False 5. False

Glossary

breed – a group of animals or plants with common ancestors.

twins – two babies born to the same mother at the same birth.